ISBN 978-0-365-39127-2
PIBN 11263751

1 MONTH OF
FREE
READING

at

www.ForgottenBooks.com

By purchasing this book you are eligible for one month membership to ForgottenBooks.com, giving you unlimited access to our entire collection of over 1,000,000 titles via our web site and mobile apps.

To claim your free month visit:

www.forgottenbooks.com/free1263751

English
Français
Deutsche
Italiano
Español
Português

www.forgottenbooks.com

Mythology Photography **Fiction**
Fishing Christianity **Art** Cooking
Essays Buddhism Freemasonry
Medicine **Biology** Music **Ancient
Egypt** Evolution Carpentry Physics
Dance Geology **Mathematics** Fitness
Shakespeare **Folklore** Yoga Marketing
Confidence Immortality Biographies
Poetry **Psychology** Witchcraft
Electronics Chemistry History **Law**
Accounting **Philosophy** Anthropology
Alchemy Drama Quantum Mechanics
Atheism Sexual Health **Ancient History**
Entrepreneurship Languages Sport
Paleontology Needlework Islam
Metaphysics Investment Archaeology
Parenting Statistics Criminology
Motivational

INTERSTATE COMMERCE COMMISSION.

REPORT OF THE CHIEF INSPECTOR OF SAFETY APPLIANCES COVERING THE INVESTIGATION OF AN ACCIDENT WHICH OCCURRED ON THE NEW YORK, NEW HAVEN & HARTFORD RAILROAD NEAR WESTERLY, R. I., ON OCTOBER 25, 1913.

APRIL 24, 1914.

To the Commission:

On October 25, 1913, there was a derailment of a passenger train on the New York, New Haven & Hartford Railroad near Westerly, R. I., which resulted in the injury of 74 passengers and 3 employees. Investigation of this accident was had in conjunction with the Public Utilities Commission of the State of Connecticut, and a public hearing was held at Providence, R. I., on October 31, 1913. As a result of the investigation of this accident I beg to submit the following report:

The derailed train was eastbound train No. 26, en route from New York, N. Y., to Boston, Mass. It consisted of three Pullman cars, all equipped with steel underframes, one smoking car, and two coaches, all of wooden construction, hauled by locomotive No. 1309. The train was in charge of Conductor Taber and Engineman Smith. Train No. 26 left Westerly at 9.25 p. m., 14 minutes late, and at about 9.30 p. m. was derailed at a point 1.6 miles east of Westerly while running at a speed estimated to have been between 30 and 35 miles per hour. Neither the engine nor the tender were derailed. With the exception of one wheel on the north rail, all of the wheels under the first Pullman car were derailed, while all the other cars in the train were derailed and came to rest on the south side of the track, some of them extending partly over the embankment.

The train broke in two between the second and third cars, the four rear cars being separated from the forward portion of the train a distance of about 150 feet. Illustration No. 1 is a view looking in a westerly direction, and shows the position of the last four cars after the derailment.

This part of the New York, New Haven & Hartford Railroad is a double-track line, and trains are operated under the controlled-manual block-signal system. Approaching the point of derailment from the west there are about 2,000 feet of tangent, all on a descending grade of about one-half of 1 per cent. The track is laid with 100-pound steel rails, 33 feet in length, single spiked to 18 or 19 untreated chestnut, oak, and pine ties, no tie-plates being used on straight track. At the point of derailment the track is on a 12-foot fill, chiefly composed of gravel. The ballast is of gravel varying from

No. 1.—View of rear portion of train looking in westerly direction.

12 to 16 inches in depth. Examination showed this track to be in good condition. It was raining at the time of the derailment.

Examination of the equipment of the derailed train showed nothing which in any way could have contributed to the derailment. Examination of the track showed that the first indication of anything wrong was a broken rail on the south side of the track. West of this broken rail there were no marks of any kind upon the rails or ties, while east of the same the ties had been cut and broken by derailed wheels, the track being torn up for a distance of about 600 feet. East of the initial point of derailment the north rail was torn out of alignment for a distance of about 12 rail lengths, while 12 successive rails on the south side were also torn up. Four of these rails on the south side were separated from each other, the bolts at the rail joints having been sheared off.

The crew of an eastbound passenger train which passed over this track less than an hour previous to the derailment testified that they felt no unevenness in the track, and that they did not notice anything which would indicate that there was anything wrong with it. Engineman Smith, of train No. 26, stated that the first thing he noticed was a slight jar or yank. He at once applied the air brakes, and on looking back saw fire flying from underneath the cars. After the accident no defects or damage of any kind were found to exist with respect to the locomotive, and he operated it through to Boston. Fireman Murphy testified that at the time of the derailment he was putting coal on the fire. He did not notice any jar from the driving wheels, being of the opinion that it came from behind the engine. The testimony of the other members of the crew shed no light as to the cause of the accident, their first intimation that there was anything wrong being the shock occasioned by the cars being derailed, coupled with the application of the air brakes.

This accident was caused by a broken rail. The investigation to determine the reason for the failure of this rail was conducted by Mr. James E. Howard, engineer physicist, whose report immediately follows:

REPORT OF ENGINEER PHYSICIST.

The broken rail which caused the derailment of train No. 26 was a C rail, 100 pounds section, open-hearth steel, of the New York, New Haven & Hartford Railroad's design, manufactured by the Bethlehem Steel Co., April, 1910, heat J–1391. It was laid in the track June 5, 1910, and therefore had been in service for a period of three years and four months at the time of derailment.

The rail had the following dimensions:

Height	6 inches.
Width of head	$2\frac{5}{8}$ inches.
Width of base	$5\frac{1}{2}$ inches.
Thickness of web	$\frac{5}{8}$ inch.
Length	33 feet.
Moment of inertia	47.18

The specifications for chemical composition governing its manufacture and the composition reported as having been furnished were:

Chemical constituents.	Called for by the specifications.	Said to have been furnished.
Carbon...........	0.70 to 0.83	0.80
Manganese.......	.60 to .90	.76
Silicon...........	.20	.183
Phosphorus......	.04	.033
Sulphur..........	....	.046

The specifications called for a drop test in which a 2,000-pound tup should be dropped from a height of 15 feet, the rail resting upon supports 3 feet apart. It was required that the rail should deflect not more than 1.45 inches on the first blow, nor upon fracture display less than 6 per cent elongation in 1 inch, or 5 per cent in 2 consecutive inches. The drop test made on this heat of steel was reported as having shown a deflection of 0.9 inch.

This rail showed very little wear as the result of its service in the track. The head retained its shape, and externally the appearance of the rail was good.

It fractured in three places at the time of derailment, at distances of 1 foot 10 inches, 4 feet 10 inches, and 6 feet 7 inches, respectively, from the leaving end. At the first and third of these places transverse fissures were disclosed measuring in diameter about $1\frac{1}{4}$ inches each. The initial line of fracture was probably that which occurred 6 feet 7 inches from the leaving end. The intermediate fracture, believed to have been a secondary one, did not have a transverse fissure.

Photograph, figure No. 2, shows the relative positions of these lines of fracture as they were viewed from the gauge side of the rail. The movement of the train on the rail was from right to left. Line of rupture CC was the first to occur, it is thought.

The train left the track southerly through the opening made by the three fragments shown on figure No. 2 and the opening made by succeeding rails east of this point. To the west of the line of rupture CC the track remained intact.

Subsequent to the derailment an additional fracture was made when removing the rail from the track, at a place about $5\frac{1}{2}$ feet west of line of rupture CC. At this place a transverse fissure $1\frac{1}{4}$ inches in diameter was displayed. The rail was then shipped to Providence, where two more transverse fissures were disclosed upon raising one end of the rail and allowing it to fall upon a concrete walk from a height of about 6 feet. These fissures measured $1\frac{3}{4}$ inches and five-eighths inch in diameter, respectively.

In all, five transverse fissures were displayed in the rail, each of which was located on the gauge side of the head. In résumé these fissures were located at the following distances from the leaving end of the rail: 1 foot 10 inches, 6 feet 7 inches, 12 feet 1 inch, 17 feet 8 inches, and 21 feet.

A second rail, from the same heat as the above, was removed from the track and its structural condition examined. This rail, branded "Bethlehem Open Hearth 100B IIII 10," was taken from the track adjacent to or near the broken rail. Both were laid at the same time, and each was exposed to the same conditions of service.

These rails were tested in part at the Bureau of Standards, while contributory work was done at the Washington Navy Yard and by the New Haven Railroad at its New Haven laboratory and at the works of the Bethlehem Steel Co.. followed by a metallographic examination by Mr. Wirt Tassin.

The report of the Bureau of Standards upon the chemical composi-

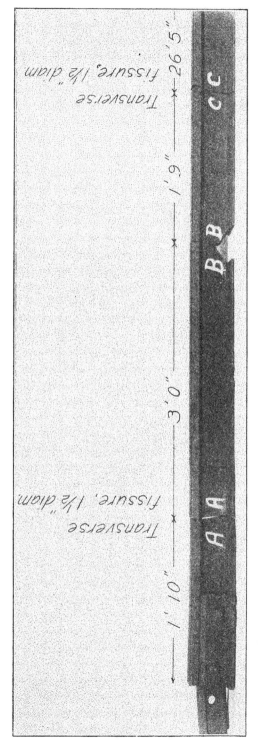

No. 2.—Appearance of east end of rail No. 1, viewed from gauge side, showing fractures which were made at the time of derailment.

tion of the steel, slag determination, metallographic examination, and tensile tests follows:

REPORT OF THE BUREAU OF STANDARDS.

CHEMICAL ANALYSIS.

In Table I are shown the results of chemical analysis of rails 1 and 2 taken at three places.

TABLE I.

Location.	Carbon.	Sulphur.	Phosphorus.	Manganese.	Silicon.	Nickel.	Oxides and slag.	Chromium.
Rail 1:								
Near running surface of head.....	0.83	0.039	0.063	0.78	0.166	0.081	0.11 .23	0.05 .05
Junction of web and head........	.82	.040	.061	.79	.164	.049	.14 .26	.04 ....
Flange of base...................	.84	.043	.063	.79	.166	.069	.17 .19	.05 ....
Rail 2:								
Near running surface of head.....	.83	.040	.058	.79	.157	.29	.07	.02
Junction of web and head........	.85	.041	.060	.79	.147	.27	.04	.03
Flange of base...................	.84	.039	.059	.80	.152	.29	.07	.02

It will be noted that the two rails, barring slag, are practically of identical composition.

The agreement between the three positions in each rail is also good, except with respect to nickel and slag in No. 1, showing no appreciable segregation, if any, of the chemical constituents. The difference as to nickel may be due to errors of analysis where such small amounts are concerned, and are probably without significance.

Attention should be called to the values reported for slag and the question of slag and oxide analysis in general. The methods used for both are very unsatisfactory, in that we have no real knowledge that they are reliable, but a good deal of reason to believe that they fail to tell us what they purport to tell. That is to say, in the case of slag we do not know if all slag is obtained by the method used, i. e., insolubility in iodine, or how much of what may be reported as slag is such. For instance, in the present case, the silica percentage in the slag found in No. 1 does not exceed 20 per cent, by actual test of several samples. This means less than 50 per cent of silicate slag, if all the silica comes from that; but if any iron silicide is included in the silica found, the slag percentage should be lowered by an indeterminate amount.

Again, this slag (ignited) carries about 9.8 per cent P_2O_5, which we may suppose to have belonged to iron phosphide. If so, and if the composition of the phosphide is Fe_3P, and if again this became converted during ignition to Fe_2O_3 and P_2O_5, we must deduct the oxygen corresponding to this change, which in the present instance would be 14.5 per cent.

Still again, the slag contained a little chromium in unknown condition. Allowing for the maximum amount of real (silicate) slag permissible as deduced from the silica percentage and of iron silicide and phosphide, there remains a large probable deficiency, which may perhaps be made up by oxide of iron or some oxide other than one of manganese, which element is not present in the slag from either rail (calcium is also absent). If the slag carried carbide and silicide of iron, the iron and silicon of these would be left after ignition as Fe_2O_3 and SiO_2. The variations in slag noted for rail 1 may be due to actual local variations in slag content or perhaps in part to uncertain analysis.

It is probable that the slag analyses are comparable for the two rails, since these are otherwise of very exactly the same composition. It is of interest to note that the rail No. 1 which failed in service had three or four times as much slag as the other rail from the same heat in the same track, suggesting a greater inherent weakness referle to this cause.

METALLOGRAPHIC EXAMINATION.

A section of rail No. 1 was cut 5 inches back of break, 12 feet from receiving end, polished and etched electrolytically by being made the anode in a bath of ammonium chloride. By this treatment the areas of segregation are shown by dark spots and streaks. Figure No. 3 shows the appearance of the section after this treatment. The web shows a considerable amount of segregation, but the metal of head and base is not seriously affected. A section of rail No. 2 treated in the same way shows a structure nearly identical with that of rail No. 1, as shown on the same figure. The amount of segregation shown by these two sections may be regarded as typical and appears to bear no intimate relation to the formation of the transverse fissures found in rail No. 1.

Sections for microscopic examination were taken from head, web and base of each of the two rails. The metal in the head was examined both from the gauge side and opposite portion. Except for an increase of grain size in the head and occasional slag threads, the structure is very uniform throughout. As near as can be judged, the microstructure of the two rails is identical. The metal consists of an intimate mixture of pearlite crystals, i. e., saturated or eutectoid steel. The method of etching used, 2 per cent nitric acid in alcohol, darkens the pearlite; the lighter appearance of many of the crystals is due to the different reflection of the light caused by the orientation of the crystals.

In the interstices between many of the crystals are areas of very coarse pearlite. Such areas are numerous and are found scattered uniformly throughout the whole mass. In these areas the two constituents of pearlite (ferrite or pure iron and cementite or carbide of iron) are in particles of sufficient size (i. e., plates) that the weakening effect upon the metal as a whole must be appreciable.

These areas can not be well represented in a photomicrograph of 100 diameters magnification. They appear as small light-colored grains.

After annealing, these interstitial pearlite areas are more pronounced and distinct.

No. 3.—Cross section of rail No. 1, on left, after etching in solution of ammonium chloride. The web shows a pronounced segregation, but the amount of foreign inclosures in the head and web is not unusual or significant. Cross section of rail No. 2, on right, was prepared in the same way as rail No. 1, and shows the same characteristics. From the macroscopic point the structure shows no significant or suggestive features.

End views A and B in figure No. 4 show the typical appearance of the transverse fissures abundant in the head of rail No. 1. No fissures were found other than on gauge side of the rail. The metal immediately adjacent to such fissures was examined in detail. Sections

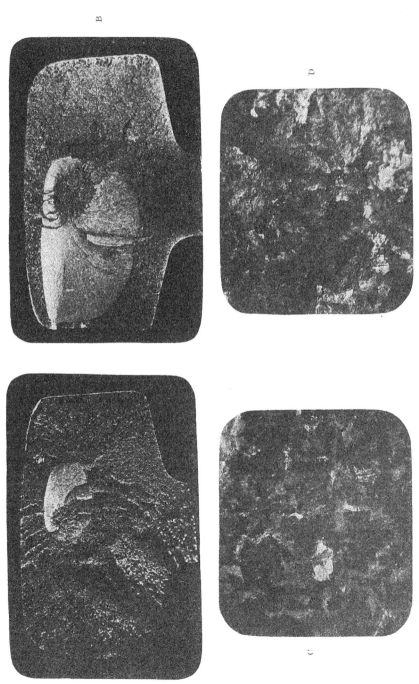

No. 4.—A. and B. Transverse fissures, from fractures of rail No. 1, the gauge side of the head being on the right and left sides respectively of these end views. C. and D. Photomicrographs, longitudinal sections, perpendicular to the faces of the fissures and just back of the same. Magnification 100 diameters. The structure appears identical with that throughout the body of the rail.

were taken perpendicular to the face of the fissure and the metal immediately back of the break examined. It was not found to differ materially in structure or constituents from that of the rail as a whole. Photomicrographs C and D show the structure of the metal just back of fissures at breaks 12 feet and 15 feet 4 inches, respectively, from the receiving end of the rail. It consists of the same mixture of pearlite crystals as is found throughout the body of the rail. There appears to be no unusual segregation of slag or foreign inclusions to be found here.

Specimens were taken from each of the two rails after annealing a section of the head of each. The occurrence and distribution of interstitial areas of coarsely laminated pearlite are here more evident. Both rails show considerable free cementite, which has coalesced as boundaries of the grains as a result of the annealing process. Cementite is the carbide of iron, F_3C, and is the hard, brittle constituent of high-carbon steels (annealed) and cast irons. It occurs here, not uniformly distributed throughout, but is more or less segregated. A sample was taken from rail No. 1 after annealing just back of the face of one of the transverse fissures. The specimen shows considerable quantities of free cementite. The estimated carbon content of such spots is over 1 per cent. The amount of cementite found in specimens from rail No. 1 was considerably more than in the samples from rail No. 2. This, however, may be only fortuitous. Before annealing, such free cementite can not be detected with certainty because it exists as isolated particles which, during annealing, coalesce to form the grain boundaries and also because the metal contains inclusions of various natures which, under the conditions, can not be differentiated from the cementite particles with certainty.

The chemical analysis indicates a carbon content of (0.82) or very slightly different from the eutectoid composition (0.85). The occurrence of so much free cementite may be attributed to the restraining action of the manganese content. The carbon is retained in the condition normally characteristic of a steel of higher carbon content. Such steel will have properties approximating those of the steel whose true carbon content is equal to the apparent content of the steel under discussion.

Tensile samples which had been cut from rail No. 2 were submitted after having received the heat treatment stated in the tabulation of the tensile tests. The structure, so far as can be distinguished under the microscope, is the same in all. They are all in the indefinite stage, sorbite, preceding the revolution into pearlite, which takes place in the critical range. The tempering temperatures were not chosen at wide enough intervals to show any decided structural change.

The critical or recalescence point of this steel was determined to be about 1250° F. The very coarse-grained structure of the rail is due to rolling at a temperature very much higher than the critical range.

The weakening effect of slag associated with coarse structure would have been lessened by rolling and finishing at lower temperatures.

No. 5.—A. View of one of the coarse interstitial pearlite areas, rail No. 1; magnification 250 diameters; etched with 2 per cent nitric acid. Such areas are very numerous throughout the body of the rail and must exert an appreciable weakening effect upon the whole.

B. Section from head of rail No. 1, after annealing; magnification 250 diameters; etching, hot sodium picrate. In the unannealed specimens the free cementite does not occur in the form of definite cell boundaries. These are caused by the coalescing of smaller particles during the annnealing process. The indefinite dark circular spots are the cut ends of the slag threads and not free cementite.

C. Photomicrograph from the base of rail No. 1, section parallel to the rolling; magnification 100 diameters; etching, 2 per cent alcohol solution of nitric acid. The structure is a jumbled mass of pearlite crystals. The long dark streaks are slag threads. There are coarse interstitial pearlite areas. (The photomicrographs of the head and web showed similar structure.)

D. Photomicrograph of tensile specimen from head of rail No. 2, after heat treatment, quenched in oil from temperature of 1,400° F. and drawn at 1,250° F. Magnification 250 diameters

All five heat-treated specimens have sorbitic structure.

SUMMARY.

1. No unusual segregation of impurities is found.
2. The structure throughout the section of the rail is very uniform.

3. The metal immediately. adjacent to the "transverse fissures" appears to be of the same nature as throughout the rest of the rail.

4. No unusual segregation of impurities or "slag" can be associated with the transverse fissures; "slag streaks," however, are present here in the usual numbers as are found throughout the rest of the head.

5. The occurrence of the interstitial areas of very coarse pearlite must have an appreciable weakening effect upon the metal as a whole.

6. The occurrence of free cementite, which in itself is a weak, brittle constituent, is a serious defect and may have a direct bearing on the formation of the "transverse fissures" developed in rails of such composition.

Tensile tests.

SPECIMENS FROM RAIL NO. 1.

Description.	Diameter.	Sectional area.	Elastic limit	Tensile strength.	Elongation.	Contraction of area.	Remarks.
	Inches.	*Sq. in.*	*Pounds per sq. in.*	*Pounds per sq. in.*	*Per cent.*	*Per cent.*	
Head of rail near initial fracture.	1.129	1.00	....	60,000	(1)	(1)	Granular.
Head of rail, outside half from west end.	1.129	1.00	60,000	78,600	(1)	(1)	Do.
Do....................	1.129	1.00	55,000	77,300	(1)	(1)	Do.
Head of rail, from west end, near center of head.	.505	.20	60,000	135,500	4.5	4.5	Do.
Head of rail, from west end, outside of head.	.505	.20	65,000	91,000	(1)	(1)	Do.
Web of rail, from west end....	.505	.20	65,000	145,800	11.0	17.0	Do.
Do.....................	.505	.20	65,000	145,500	12.0	17.0	Do.
Base of rail, from west end....	.505	.20	65,000	146,300	10.0	9.0	Do.
Do.....................	.505	.20	65,000	148,700	11.0	15.0	Do.
Do.....................	.505	.20	65,000	146,000	11.0	15.0	Do.
Do.....................	.505	.20	65,000	142,000	10.0	14.5	Do.

[1] Inappreciable.

SPECIMENS FROM RAIL NO. 2.

Description.	Diameter.	Sectional area.	Elastic limit	Tensile strength.	Elongation.	Contraction of area.	Remarks.
Head of rail, middle of.......	1.129	1.00	....	49,800	(1)	(1)	Granular.
Gauge side of head..........	.505	.20	65,000	144,400	9.0	12.5	Do.
Center of head..............	.505	.20	65,000	77,500	1.0	.5	Do.
Outside of head.............	.505	.20	70,000	140,800	3.0	3.5	Do.
Web of rail.................	.505	.20	65,000	147,000	10.5	13.5	Do.
Base of rail................	.505	.20	70,000	149,000	11.0	15.5	Do.
Do.....................	.505	.20	70,000	150,400	10.0	15.0	Do.
Head of rail, middle of, annealed.	1.128	1.00	60,000	66,300	(1)	(1)	(2)
Head forged down and annealed.	1.129	1.00	....	95,400	1.6	(1)	(3)
Head forged down to 1-inch diameter, annealed, reheated to 1,400° F., and quenched in oil, then drawn as described:							
Drawn at 1,000° F........	.505	.20	105,000	150,800	15.5	37.5	Silky.
1,080° F........	.505	.20	90,000	148,000	17.0	42.0	Do.
1,150° F........	.505	.20	95,000	148,900	17.0	41.0	Do.
1,200° F........	.505	.20	95,000	150,850	18.0	38.0	Do.
1,250° F........	.505	.20	100,000	144,000	16.0	37.0	Do.

[1] Inappreciable.   [2] Granular blue-black spot ⅜ by ¼ inch.   [3] Granular; broke in head at root of thread.

DROP TESTS AT THE WORKS OF THE MARYLAND STEEL CO.

Drop tests were made at the works of the Maryland Steel Co. on one piece of rail No. 1 and two pieces of rail No. 2. Prior to these tests the second rail was broken in several places by bending loads applied in the testing machine at the Bureau of Standards, but none of the fractured surfaces showed transverse fissures. The drop tests were made with the rail sections head up, supports 3 feet apart, 2,000 pounds tup, height of fall 15 feet. The rails were tested at a temperature of 90 degrees.

Rail No. 1 sustained the first blow without rupture, showing a deflection of 0.84 inch. It broke on the second blow, with an extension of the metal of 6 per cent. The sections from the second rail each broke on the first blow, neither developing appreciable extension. The appearance of these pieces is shown by figure No. 6. The upper rail in the cut represents No. 1. The middle and lower sections were those from rail No. 2.

It will be noted that rail No. 1, which failed in the track and caused the derailment of train No. 26, successfully passed the prescribed drop test, displaying an extension of 6 per cent as required, while the second rail, which failed to meet the drop test, did not fail in the track.

The fracture near the bolt holes of one section of rail No. 2, was secondary, occurring when this fragment struck the bed of the drop testing machine succeeding the blow of the tup.

The section of rail No. 2, represented by the lower figure of the cut, did not fracture under the place directly struck by the tup, but sheared out a fragment 22 inches long, symmetrical with the supports.

In prescribing only 6 per cent extension of the metal under the drop test, as an index of the ultimate ductility of the steel, it may be said that such extension, developed as it is under transverse stress, is not far above the zero limit of ductility. In the milder grades of steel the extension under transverse stresses commonly exceeds many times that witnessed in the tensile tests of the metal. In these two rails the reverse was true. The 2-inch tensile specimens from the bases of these rails showed 10 and 11 per cent extension, which, equated for specimens of greater length of uniform section, would still be more than the extension in the drop test of rail No. 1, which was 6 per cent, while No. 2 failed with zero extension.

The mechanical work required to rupture steels of high elastic limit but incapable of permanent set is small compared with the work required to rupture mild steels which display the extension usual in structural steels. If the indications of the present tests are confirmed and it is found that rails normally of limited extension tend to fail under the effect of rapidly applied loads without appreciable set,

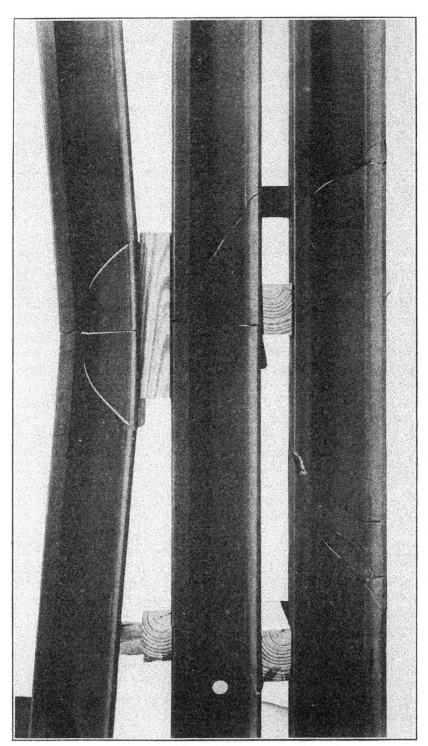

No. 6.—Appearance of sections of rail No. 1 and No. 2 after drop test.   Upper figure of cut represents section of rail No. 1.   Intermediate and lower figures represent sections of rail No. 2.

this feature in the use of hard steels will demand early considera-
tion, in which striking velocities and temperatures of the rail should
be included.

The striking velocity of the tup with 15 feet height of drop, as
prescribed in current specifications, is a low velocity compared with
ordinary train speeds.

TESTS BY THE NEW YORK, NEW HAVEN & HARTFORD RAILROAD CO.

The New York, New Haven & Hartford Railroad Co. conducted an
examination, which comprised tensile tests, chemical analyses, metallographic examination, and drop tests, the later at the works of the
Bethlehem Steel Co., other parts at its New Haven laboraory.   In
addition to supplying data confirming the information from other
sources, this examination resulted in showing the presence of a considerable number of incipient fissures in the head of rail No. 1, at
places where the ordinary manifestations of transverse fissures were
not in evidence.   That is, the fissures had not reached the advanced
stage in which they would ordinarily be detected by visual inspection.
Deep etching, with nitric and hydrochloric acids, developed short
transverse cracks or incipient fissures which were thus rendered
plainly visible to the eye.   They ranged in length from a few hun-
dredths of an inch to three-eighths inch.   So far as could be judged
the zone of greatest structural disturbance was in the head over the
gauge side of the web and toward the gauge side.   The etching being
very deep not unlikely brought into view fissures which originally
were less easily discerned.

A group of these incipient fissures is shown in figure No. 7, (a) and
(b).   The fissures are here shown about natural size, the same group
appearing in both (a) and (b).   In the latter they are partially ob-
literated by rough polishing the surface while removing some longitudinal scratches and introducing others.   The location of the zone in
which these fissures were found leads to the inference that we are here
examining the same class of phenomenon witnessed in the larger
transverse fissures, which ultimately result in the complete fracture
of the rail.   If such is the case, structurally, a more general disinte-
grating effect has been brought about than indicated by the display of
transverse fissures in the rail—that is, influences which tend toward
the formation of transverse fissures are not localized at those places
only where fissures have reached an advanced stage of development.

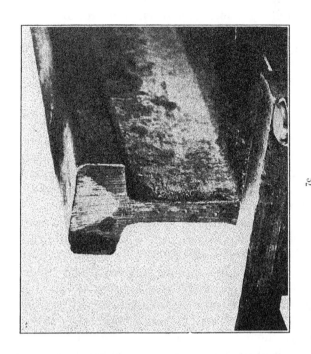

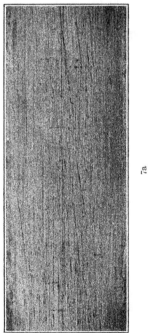

No. 7.—Incipient transverse fissures in head, near east end, of rail No. 1, as shown on a longitudinal section, cutting head in a vertical plane. Fissues shown natural size, same fissues appearing both in 7a and 7b. End view of head of rail of another series, 7c, showing wedge-shaped zone of metal disturbed by high wheel loads, in reciprocating test.

Other rails were taken from service by the New York, New Haven & Hartford Railroad Co., and used in this examination. One, an **E** rail, designated as No. 10, was taken from the track under the supposition that it belonged to the same heat of steel as No. 1, which caused the derailment. Analysis, however, showed that it came from another heat. Check analysis at the steel works showed the following composition:

Carbon, 0.83. Manganese, 0.49. Phosphorus, 0.020. Sulphur, 0.052.

This rail did not break, as others had, when lifted with a magnet and dropped bodily from a height of $9\frac{1}{2}$ feet. When ruptured under the regular drop test, head up, three breaks were made, each of which were reported as having shown clean metal, free from transverse fissures. A test of the metal from the gauge side of the head showed a tensile strength of 122,000 pounds per square inch, with a contraction of area of 20 per cent.

The Bethlehem Steel Co. in reporting upon the examination of the material pertaining to this inquiry state:

As in all other cases of rails developing transverse fissures in the head, no segregation of any kind was found in the rails to account for this defect. There was no difference found between the microstructure of the core of the fissures, the bright parts of the fissure, or any other part of the rail. All fissures occurred on the gauge side of the head of the rails.

A rail rolled in the same year and month and branded the same as rail No. 1, but which had not been used in the track, was taken from a spare rail post and cut up for examination. The chemical composition of this rail was found to be:

Carbon, 0.84. Manganese, 0.87. Phosphorus, 0.037. Sulphur, 0.025.

A tensile specimen of 1 square inch sectional area and 10 inches long, taken from the center of the head, gave the following results:

Elastic limit, 60,000 pounds per square inch.
Tensile strength, 148,000 pounds per square inch.
Elongation, 6.3 per cent.
Contraction of area, 6.9 per cent.
Appearance of fracture, granular.

———

Sections of several rails included in this inquiry were cut out for photographic purposes, metallographic examination and check determination of the carbon at the Washington Navy Yard. Metallographic examination was made by Mr. Wirt Tassin, with the assistance of Mr. Paul E. McKinney.

REPORT OF METALLOGRAPHIC EXAMINATION BY MR. WIRT TASSIN.

The material examined included two specimens identified as "Westerly Rail No. 1, which failed in the track," with five specimens identified as "Rail No. 10, which had been in service, but did not fail in the track."

All work done was metallographic.

WESTERLY RAIL NO. 1, WHICH FAILED IN THE TRACK.

Figures 8a, 8b, and 8c show macroscopic transverse fissures in the rail head, as seen at a magnification of 8, on a specimen taken from the "west end" of the rail. They are plainly visible to the unaided eye and are not associated with slag, sulphide, or oxide areas, nor are they accompanied by segregations of any kind.

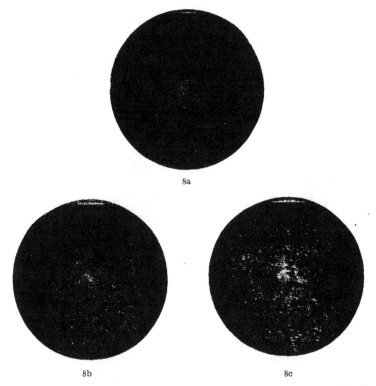

8a

8b   8c

No. 8 —Macroscopic transverse fissures in the head of rail No. 1, near its west end. Magnification 8 diameters. Fissures are not associated with slag, sulphide, or oxide areas, nor accompanied by segregations of any kind.

Figure 9a, at a magnification of 315, using a B. and L. 8 mm. objective and a 15x eyepiece, shows a transverse microscopic fissure as seen on a longitudinal section cut from the "east end" of the rail and located at about the upper center of the head. The fissure could be readily traced for a distance of 4 mm., 0.15748 inch.

Figure 9b, at a similar magnification, is another transverse fissure whose length could be traced for 11 mm., 0.43307 inch.

Figure 9c, at a magnification of 315, shows a typical area of incipient fissures.

9a

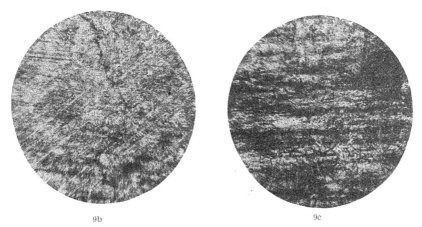

9b   9c

No. 9. Microscopic transverse fissures from head of rail No. 1, near its east end; magnification 315
    diameters.
9a was traced for a length of 0.157 inch.
9b for a length of 0.433 inch.
9c shows a typical area of incipient fissures. These fissures are not associated with and have no rela-
    tion to any areas of sulphide, slag, or other inclusions.

In each instance it will be noted that these fissures are not associated with and have no relation to any areas of sulphide, slag, and other inclusions. This is further shown in figures 10a, 10b, and 10c, which are at the same magnification and show seams of such inclusions with a complete freedom from microscopic or incipient fissures.

The general structure of the rail is sorbitic. There is little or no lamellar pearlite, no free cementite or ferrite. The sulphide, slag, and oxide areas are small and sparingly distributed. There are no segregations.

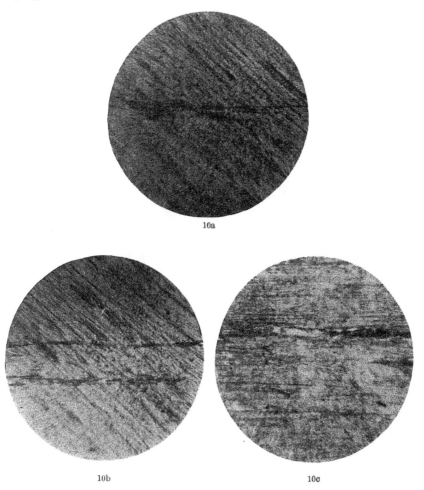

10a

10b   10c

No. 10. Showing slag and sulphide seams in head of rail No. 1. Magnification 315 diameters. At these seams there was complete freedom from microscopic or incipient fissures.

RAIL NO. 10, WHICH HAD BEEN IN SERVICE BUT DID NOT FAIL IN THE TRACK.

Figure 11 is a sketch showing the location in the rail of the sections examined.

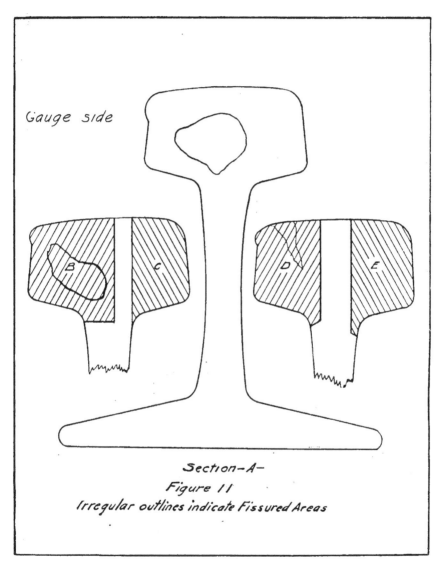

No. 11.—Cross section of rail No. 10, showing manner of cutting up for examination, and location of fissured areas found therein.

Figure 12a, magnification 315, shows incipient transverse fissures in the center of the head of section A.

Figure 12b, magnification and real field as in figure 12a, is from the lower center of the head of section A and shows incipient fissures.

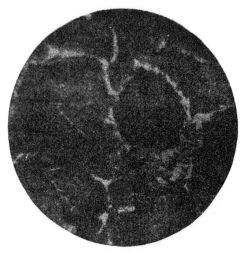

12a

12b .

No. 12. Incipient fissures found in the head of rail No. 10.
12a represents fissures found in the center of the head, in section of rail marked "A."
12b, fissures found in the lower center of the head. Magnification, 315 diameters.

Figures 13a and 13b, magnification 315, show the smallness of the seams of slag and sulphide as found in section A.

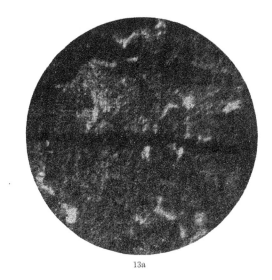

13a

13b

No. 13. From head of rail No. 10, section marked "A,"-showing the smallness of the seams of slag and sulphide.   Magnification, 315 diameters.

Figures 14a, 14b, and 14c, magnification 315 diameters, show typical incipient fissures as found in specimen B. Figures 15a and 15b show analogous fields in section D.

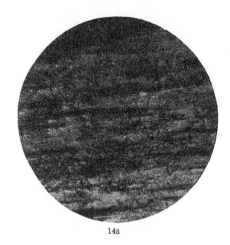

14a

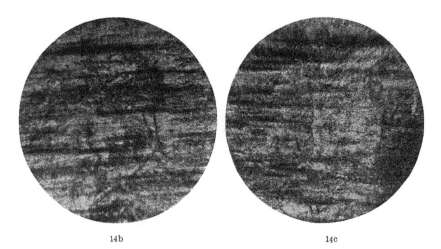

14b   14c

No. 14. Typical incipient fissures found in head of rail No. 10, section marked "B." Magnification, 315 diameters.

Figure 15c shows one of the largest of the sulphide-slag areas and illustrates the smallness of these inclusions.

Specimens C and E (see fig. 11) show no incipient fissures and no abnormalities of structure.

Plotting the fissured areas as seen in figure 11, it will be noted that they are practically limited to the gauge side at or near the center of the head and are in the regions immediately affected by the wheel loads.

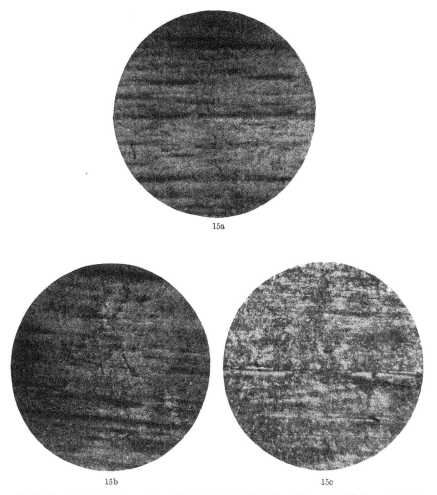

No. 15. Typical incipient fissures found in the head of rail No. 10, section marked "D," represented by 15a and 15b.   One of the largest sulphide-slag areas of rail No. 10, illustrating the smallness of these inclusions is represented by 15c. Magnification, 315 diameters.

The opinion is advanced that these fissures are set up by these loads, and in support of this the statement is made that new rails which had not been in the track and which have as high and even higher carbon content do not show under the microscope similar fissured areas.

CONCLUSIONS.

The Westerly rail No. 1 shows nothing in its structure to indicate any abnormalities in the mill practice. The rail is comparatively free from slag, sulphide, and oxide areas. None of the areas showing macroscopic or microscopic fissures can be correlated with areas containing sulphide, slag, oxide, or segregations.

Rail No. 10 shows nothing in its structure to indicate any abnormalities in the mill practice. Slag, sulphide, and oxide areas are very sparingly distributed and are relatively small. Numerous incipient fissures are present and are located in the head on the gauge side.

The general structure of the two rails is such as to warrant the statement that they are of a carbon content that will not afford a toughness and ductility comparable with that of a properly treated rail of a lower carbon content, hence the transverse fissures.

To remedy this condition two courses are open—fix the upper carbon limit and prescribe the mill treatment which will insure the maximum toughness and ductility with a sufficient strength, or reduce the wheel loads. The former plan is to be preferred.

The metallographic examination by Mr. Tassin completed the present examination of the material pertaining to this derailment.

Concerning the prevalence of transverse fissures in steel rails, of a macroscopic order, not referring at this time to those microscopic, of such dimensions as are menacing to the safety of railway travel, they are believed to be numerous. Forty-six transverse fissures of recent occurrence have been reported in 32 rails. Instances have been reported in which five transverse fissures have been found in the same rail within the limits of 3 feet. They are present in both open-hearth and Bessemer steels. They are not confined to the product of the rail mills of one section of the country. They occur over ties and between ties, near the ends of the rails and along the middle of their lengths, on tangents and in the upper as well as the lower rails on curves. But one general remark can be made—they persistently appear on the gauge side of the head.

In their maximum state of development they have been witnessed in 100-pound rails having attained a superficial area of 3.3 square inches, leaving practically only the web and the base intact. Our investigation shows without question that these hidden fissures in some rails reach such a state of development before discovery as to destroy nearly the entire head of the rail, therefore it is not reassuring that other rails of similar composition, working under similar conditions of service, are not free from these interior defects. The continuance of conditions which have resulted in derailments, attended with loss of life and injury to passengers and employees, places a great responsibility upon all who can in any manner aid in the inauguration of measures which will tend immediately to ameliorate these grave conditions.

Reference has been made in earlier derailment reports to causes which are believed to be contributory to the formation and development of transverse fissures. Data have since been gathered, some of which are embodied herein, illustrating the probable sequence of events which attend their formation, from the results of which the contributory causes more clearly admit of recognition. Tracing the fissures in the reverse order of their development, they are followed down from the larger ones of rails broken in the track having darkened oxidized surfaces to those of an earlier stage of development, of smaller diameter with bright silvery surfaces, which have not yet extended to the outer surface of the rail and therefore retain the brightness of freshly fractured surfaces.

Next earlier appear fine cracks, not easily discernible to the eye until the steel has been etched, preceded by other cracks still finer of a microscopic order, but of measurable length to the eye if the cracks were of sufficient width to be seen, eventually leading to the detection of microfissures and the fragmentary or partial separation of the microconstituents. These indications follow each other in such order that they appear to belong together, representing different stages in the loss of structural integrity and destruction of the metal of the rail.

Such are the indications which pertain to rails which have been in service. They have been looked for, but not found, in steel which has not been in service, of similar composition and concurrent manufacture. As the evidence stands the formation of such fissures seems attributable to track conditions and to those stresses which reach a maximum in the metal of the rail at the head.

Steel of the composition of these rails in inherently a strong steel. Music wire is drawn from steel of 0.85 carbon and in the form of fine wire displays great strength. Records of such wire show a tensile strength exceeding 450,000 pounds per square inch. This grade of steel, in the form of hot-rolled, cylindrical bars, also, will endure repeated stresses applied many million times, as high as 40,000 pounds per square inch. Furthermore, in these rails the steel has displayed, in specimens from unaffected parts of the cross section, tensile strength in round numbers ranging from 140,000 to 150,000 pounds per square inch. The strength possessed by steel of this carbon content shows that in dealing with rails of such composition we have a metal which in its original state has high physical properties.

The spontaneous rupture of steels from internal causes occasionally occurs with metal hardened by heating and quenching. Intense internal strains are set up by sudden quenching which, if not ameliorated by drawing the temper, may occasion spontaneous rupture. Coils of hard-drawn wire have shown fractured ends developed a few hours after drawing. Castings also have failed spontaneously. These examples, however, refer to other classes of material than rails, while internal tendencies, if any of special account exist in steel rails due to casting or rolling conditions, have not so far as known materially contributed toward rupture. Internal strains are left in steel rails as in all hot rolled and naturally cooled steel, modified by shape, weight of section, and rate of cooling, and higher strains may exist in hard steels over those of lower grades in composition. These features are all known to exist, and while in a strict sense they are not negligible factors in the use of steels, nevertheless so far as pertains to the formation of transverse fissures in steel rails they are not held to be vital.

In reference to the relative strength of steel at the interior and exterior of the shapes in the different passes of the bloom and rail mills: The metal at the interior has been found lower in tensile strength than that near the surface, in the early passes. However, when the finished rail is reached these differences are inconspicuous, but with the thinner sections of the web and base commonly showing higher strength than the heavier section of the head. In this inquiry the strength of metal at the center of the head of an unused rail, one taken from a spare rail post, was found to be 148,000 pounds per square inch. This indicated normal strength, showing no inherent weakness in the metal of that portion of the head in which the weakened steel of rails, taken from the track, has been found.

Gagging is an operation common in all rail mills. Necessarily the elastic limit of the metal must be exceeded in this operation, otherwise no permanent straightening would result, and since the elastic limit must be exceeded, it follows that the higher the elastic limit of the steel the greater will be the overstraining force required to effect the straightening.

The actual bending force required for the purpose obviously may be modified by regulating the distance between the supports of the gagging press, but the resultant longitudinal strain in the rail must in any event be sufficient to balance the internal strains which tend to cause the rail to return to its original bent shape. These internal strains will be greater in steels of high physical properties than in mild grades of metal. In prescribing hard steel for rails these severe internal conditions will necessarily be encountered.

Significance would attach to these several features in searching for causes leading to the development of transverse fissures if intimate relations were found to exist between them and such defects. While not entirely disassociated, their influence and mutual relations seem remotely connected. Evidence connecting the formation of transverse fissures with these primitive conditions is less direct than that which attaches to track conditions.

In reviewing the results of this inquiry it will be noted that the tensile strength of a rail of substantially the grade of metal represented in rail No. 1 was 148,000 pounds per square inch. This strength was found at the center of the head of a rail which had not been in service, but one which was rolled in the same year and month and by the same manufacturer as rail No. 1. Microscopically, no fissures were found in this unused rail.

Rail No. 1 and its companion No. 2 showed substantially the same strength in each of their webs and bases, closely approached by the metal in some parts of their heads. . Metal from the head of No. 1, forged down and heat treated, gave correspondingly high results. The normal strength of the steel, as indicated in these tests, ranged from 140,000 to 150,000 pounds per square inch. In general, however, the metal in the heads of both rails, taken in the condition the rail came from the track, showed impaired strength. A tensile specimen from the head of rail No. 1 fractured at 60,000 pounds per square inch, while a corresponding specimen from No. 2 failed at 49,800 pounds per square inch. An annealed specimen from No. 2 had a tensile strength of 66,300 pounds per square inch. The latter ruptured at a preexisting defect, an interior fissure of small size, which was shown by an oxidized spot at the circumference of the test piece. Tests conducted by the New Haven Railroad on these features of the case furnished corroborative results. The metal in the head of each rail was in a weakened condition.

The metallographic examination of No. 1 showed the presence of incipient fissures and a condition of structural unsoundness existing in different parts of the length of the rail. This unsoundness affected the metal in the center of the head and on the gauge side. A generally shattered state characterized the metal of certain parts of the head, while in other parts of the cross section of the rail the steel was intact.

Although rail No. 1 had developed transverse fissures and had broken in the track, while No. 2 had not experienced such extreme vicissitudes, nevertheless the weakened condition of the metal in the head of each was about the same. Had rail No. 2 been kept in the track it is believed that it would eventually have developed the same kind of fissures as shown by No. 1. The zones in which this state of weakness was found where located in the same places in each rail and closely coincided with the zone of disturbed metal which had been experimentally made in reciprocating tests of rails under high wheel pressures.

Another rail of this series was microscopically examined, one taken from the track after the same period of service as rail No. 1, from which it differed in having lower phosphorus and manganese. Its phosphorus content was 0.020; manganese, 0.49. This rail showed similar incipient fissures to those of No. 1.

Loss in tensile strength, the display of macroscopic and microscopic fissures, and the development of transverse fissures, all seem to be associated phenomena, and from the location of the affected metal seem traceable to the action of wheel loads for their development. No other cause is recognized as being present to which their formation could be ascribed.

The general condition of the steel in these tests outside of the affected zone was good and possessed of normal strength. The metallographic structure was the same at the fissures as in other portions of the rail. There were no slag inclusions or indications of the presence of slag at the transverse fissures where the rail fractured, nor was slag found at the microfissures, of which there were many examples in the different rails. Slag inclusions did not, so far as can be ascertained, locate either the places of incipient fissures or the larger transverse fissures of the broken rail. Since such inclusions would be acicular in shape, in size not materially detracting from the sectional area of the steel, and drawn out parallel to the length of the rail, they would not be expected to exert any appreciable influence on the strength of the rail against longitudinal stresses. A number of slag streaks were specially examined for the purpose of ascertaining whether incipient fissures had their origin at such streaks, but no fissures were found associated with them. Metallographically normal mill practice is indicated.

In conclusion it appears—

That the derailment of train No. 26 was due to a broken rail.

That the rail fractured under this train by reason of the presence of transverse fissures in its head, one of which was located 6 feet 7 inches from the leaving end and is believed to have been the place of initial fracture.

That five transverse fissures were found in the rail ranging in diameter from five-eighths inch to 1¾ inches.

That fissures of lesser extent were present in both this rail and a companion rail of the same heat, each of which had had the same term of service in the track.

That the metal in the heads of these rails was in a weakened state in certain affected zones, the affected zones being located near the center of the head and toward the gauge side.

That the weakened and structurally impaired metal shown in the tensile tests was confirmed by the metallographic examination.

That the steel in other parts of the rail was structurally sound and possessed of normal strength.

That no slag inclusions were associated with the transverse fissures nor with the microfissures.

That fissures were present in different stages of. development, associated with each other apparently as to a common cause, the microscopic examination indicating that such fissures were located in metal otherwise structurally sound.

That microscopic fissures were present in certain other used rails, which had not yet developed full-sized transverse fissures, furnishing evidence that such rails were approaching a state in which full-sized fissures would eventually be formed.

That the proximate causes to which the transverse fissures in broken rail No. 1 are ascribed are high wheel loads with their attending strains, evidence of other causes not having been found.

That testimony is to the effect that a considerable number of rail failures have recently occurred by reason of the presence of transverse fissures.

That conditions which were precursors to the formation of transverse fissures in broken rail No. 1 exist in other rails now in service.

That the presence in the track and continued use of rails of the same or similar composition to rail No. 1 and exposed to the same service conditions is a source of danger.

That evidence acquired indicates that transverse fissures may be and are formed through the action of high wheel loads upon hard steel rails.

It is manifestly evident from the above report and conclusions of Mr. Howard, which are concurred in, that in this type of rail failures there is presented a serious situation. Rail failures of other types have been the cause of many accidents.

The figures contained in the following table are taken from the monthly accident reports made to the Commission by the railroads and show the number of derailments caused by broken rails which have occurred yearly since July 1, 1901, together with the casualties and monetary loss resulting therefrom:

Year.	Number of accidents.	Killed.	Injured.	Damage, including cost of clearing wreckage.
1902	78	5	207	$128,769
1903	150	12	204	166,140
1904	176	9	139	157,682
1905	201	4	465	257,519
1906	220	7	635	254,862
1907	308	12	699	284,675
1908	238	16	433	296,327
1909	196	5	498	191,842
1910	243	24	369	293,899
1911	249	12	463	292,749
1912	363	52	1,065	511,778
1913	340	17	827	401,551
Total	2,762	175	6,004	3,237,793

This enumeration of casualties refers to a period the greater part of which elapsed before rail failures by transverse fissures were known or had become so prevalent as they now appear. The tabulated results, however, emphasize the importance which the subject of safety in rails has assumed. To those elements of danger existing in the past is now added this type of failure shown in the development of interior fissures. On account of the insidious character of these fissures and the fact that they are progressive in their development, and so far as is known no system of inspection has been found that will detect them until they have reached the surface of the rail, make it extremely difficult to suggest any preventive against future accidents of this character.

Although as noted in previous reports dealing with rails failing on account of transverse fissures, it seems apparent that a remedy lies in the diminishing of wheel pressures and the lowering of direct compressive bending and shearing stresses.

From the constant increase in rail breakages occurring from this new type of failure, it would appear that the danger zone in the use of steel rails as at present manufactured has been reached, since the study of the rails here under discussion appears to indicate that transverse fissures are the direct result of high-wheel pressures acting upon hard steel. A complete investigation of rail, track, and wheel load conditions for the purpose of determining the effect thereon of the recent types of locomotives and cars, with their greatly increased wheel loads, should be undertaken for the purpose of scientifically determining this matter and ascertaining a remedy.

Respectfully submitted.

H. W. BELNAP,
Chief Inspector of Safety Appliances.

O

WASHINGTON : GOVERNMENT PRINTING OFFICE : 1914

CPSIA information can be obtained
at www.ICGtesting.com
Printed in the USA
LVHW080809051218
599325LV00003B/315/P

9 780365 391272